Fabiana Aparecida Leite Bomfim
Vânia H. de Almeida

The mathematical knowledge of miners in the Matupá region

Fabiana Aparecida Leite Bomfim
Vânia H. de Almeida

The mathematical knowledge of miners in the Matupá region

Ethnomathematical research

ScienciaScripts

Imprint

Any brand names and product names mentioned in this book are subject to trademark, brand or patent protection and are trademarks or registered trademarks of their respective holders. The use of brand names, product names, common names, trade names, product descriptions etc. even without a particular marking in this work is in no way to be construed to mean that such names may be regarded as unrestricted in respect of trademark and brand protection legislation and could thus be used by anyone.

Cover image: www.ingimage.com

This book is a translation from the original published under ISBN 978-613-9-61479-0.

Publisher:
Sciencia Scripts
is a trademark of
Dodo Books Indian Ocean Ltd. and OmniScriptum S.R.L publishing group

120 High Road, East Finchley, London, N2 9ED, United Kingdom
Str. Armeneasca 28/1, office 1, Chisinau MD-2012, Republic of Moldova, Europe
Printed at: see last page
ISBN: 978-620-7-75309-3

Copyright © Fabiana Aparecida Leite Bomfim, Vânia H. de Almeida
Copyright © 2024 Dodo Books Indian Ocean Ltd. and OmniScriptum S.R.L publishing group

CONTENTS

THE MATHEMATICAL **KNOWLEDGE** of gold MINERS IN THE MATUPA REGION

I dedicate this work to my husband, Tassio de Souza Bomfim, the person I love most in life. Thank you for your affection, your patience and your ability to bring me peace throughout each semester.

To the teacher and course coordinator Vânia Horner de Almeida for her conviviality, support, understanding and friendship.

"We know very little about what we are, and even less about what we can be."

Lord Byron

SUMMARY

LEITE, Fabiana Aparecida. **"The mathematical knowledge of miners in the Matupâ region".** Year, 2018 f. Course Conclusion Work (Full Degree in Mathematics) - Mato Grosso State University, Matupâ.

This paper describes an investigation based on EthnoMathematics. The approach taken, from its remarkable and comprehensive perspective, seeks to demonstrate the ways in which a more productive and lasting teaching practice can be implemented. The aim is to reveal the mathematical knowledge retained by the miners who live on the outskirts of the municipality of Matupâ-MT, particularly those who did not attend school until the end of secondary school. To this end, a survey was carried out through interviews, identifying the level of schooling and skills possessed by these workers, how they perform their mathematical calculations and also the relationship with what they earn. The data was collected from students in the third year of secondary school at a public school in the municipality of Matupâ, whose subjects had the opportunity to experience praxis in math classes. It also highlights the importance of the application of EthnoMathematics in the social and economic sphere of the region, also considering from an academic point of view, the fact of low schooling, but the enormous body of knowledge obtained in the relationship between theory and practice arising solely and exclusively through the profession.

Keywords : Mathematics Education, Informal Mathematics, EthnoMathematics.

1. INTRODUCTION

The reasons behind the choice of the theme "The mathematical knowledge of gold miners in the Matupà region" are due to the importance that gold mining has had and still has for the economic activities of the municipality and the region, we understand that the project is relevant because it brings to society a little of the mathematical culture of gold miners and, It also helps to bring the context of the activities carried out in this environment into the classroom, to cover the process of mining from extraction to the sale of gold, to investigate why this practice, which arises from necessity even in times of scarcity, has become the only source of income for countless families in the north of Mato Grosso.

The relevance of working with this area of knowledge is that it reveals the contribution that the EthnoMathematics trend makes to teaching and learning processes, by seeking to include the notion of contemporary mathematics in the common life of students with mathematics, thereby reducing the difficulties presented by them in teaching this subject. It also highlights the importance of research for society - understanding that mathematical knowledge is the product of human work. In this sense, educators - by appropriating this area of knowledge in the process of teaching and learning contextualized mathematics - broaden students' horizons in a meaningful way.

In this exploration, we will approach the research as an idea that sought to get to know the practice of Mathematics in this group, connecting the social environment in which it takes place, initiating a trajectory for reproductions in the teaching of

Mathematics, as analysis, research and pedagogical application.

We need to take a closer look at the diversity that Mathematics offers us when treated from a socio-cultural perspective, and how few people still have access to it. That's why we hope to reveal EthnoMathematics in teaching and learning, by encouraging and developing creativity, to take this idea to school, to pedagogically plan the trend so that more work like this takes place in the classroom.

Math classes are often seen as dull and tiresome by students, since they don't really know the value of this area of knowledge in building citizenship, its great importance in the job market, in shaping intellectual abilities and speeding up their reasoning. So, one of the alternatives for these classes to stop being tiresome would be to show students the importance of Mathematics, approaching the content with facts from the daily lives of high school students, valuing the reality of each one. For this to happen, EthnoMathematics is a great way of helping to revise the content, looking for methodologies that are compatible and effective with the education that society wants.

When students feel valued, they also learn to value their origins and come to believe that they can change their realities if need be and build their own history.

> [...] From an educational point of view, it seeks to understand thought processes, ways of explaining, understanding and acting in reality, within the cultural context of each individual. EthnoMathematics seeks to start from reality and arrive at pedagogical action in a natural way, through a cognitive approach with a strong cultural foundation (BRASIL, 2001, p. 23).

EthnoMathematics seeks to understand the mathematics of different cultural groups and working in the classroom from this perspective enables students to understand this empirical mathematics and relate it to the formal mathematics proposed in the school curriculum. Based on this, our guiding question is: What mathematical knowledge do gold miners have and how do they use it in their gold mining activities in the Matupà region?

The general objective is to identify the mathematical knowledge that gold miners possess and use in gold extraction and marketing activities in the Matupà region. The specific objectives are to highlight Mathematical Knowledge through EthnoMathematics, showing its importance for social groups; to investigate the mining process, the practices used by the prospectors, their experiences, their knowledge and their strategies in carrying out the activities and to apply a didactic sequence on the subject in the classroom, so that the students also take part in the work.

The purpose of this monograph is to present the results of a study carried out with miners living in the municipality of Matupâ-MT in 2017. It falls into the category of research, with the intention of mapping, debating and evaluating the results of the EthnoMathematical investigation. The significance, therefore, lies in the assimilation of the cultural activity present, in other words, a contemplative observation of the particularities that are established within this group.

Thus, in a unique and exclusive way, the project was applied to the mathematics content of students in the 3rd year of secondary school at the Jardim das Flores State

School. With this totally different way of learning mathematics, the students had the opportunity to acquire not only the basics of the subject, but also to take part in an experience that values the culture of the place where they live. In this way, the students, the subjects of this research, ended up being part of the history of each of the miners interviewed.

In short, the project was accepted by the school's teachers and management team, with whom a pedagogical proposal was developed with the intention of promoting the direct participation of the students. After studying basic concepts in class, they were taken out into the field to do research, using qualitative interviews of a free-ranging nature, i.e. with the aim of promoting an informal dialogue that would allow them to obtain data relevant to the study. In all, they were able to talk to nine miners from the region, who were either parents or relatives of the students. This activity, being in a different format, allowed them to transform oral language into academic language.

The following steps were taken to carry out the proposed activities:

S Drawing up a lesson plan;

S Official communication and acceptance of the proposal to the school board;

J Presentation of the proposal to the students;

J Studies of percentage, area, perimeter and volume;

J Presentation of field research for data collection;

J Analysis and description of the results obtained.

The following are previews of each chapter of this work:

Specific objective 01: To highlight the EthnoMathematical trend;

Preview of chapter 01: The topic begins with the conceptualization of EthnoMathematics, as well as highlighting its properties, popularity and results. It then looks at the evolution of the trend in Brazil and around the world, focusing on Ubiratan D'ambrósio who was one of the mathematicians who helped to reformulate and popularize this program.

Specific objective 02: Describe the beginnings of mining in the municipality of Matupà;

Preview of chapter 02: In the second chapter, the text gives a little background to the history of the municipality and the first prospectors who arrived in the region. It then reveals how the gold extraction processes are carried out and the conditions experienced by the workers.

Specific objective 03: To present how the project succeeded in the school environment;

Preface to chapter 03: The final chapter provides a complete analysis of the moments when the students were able to take part in the research; how they performed; the results obtained through this connection between formal knowledge and popular knowledge.

2. ETHNOMATHEMATICS: ITS ORIGINS AND PURPOSES

This chapter will discuss the meaning of the term EthnoMathematics, a trend revealed in Brazil by the mathematician Ubiratan D'Ambròsio and praised by countless other mathematicians, the way in which mathematics has changed from ancient times to the present day. The aim is also to evaluate the progress of EthnoMathematics and its analogies with a view to using it as a teaching model.

2.1 DEFINITION

Mathematics is, and always has been, one of society's most important tools. Improving basic mathematical concepts and processes helps to shape future citizens who will engage in the world of work, continue their school life and participate in social and cultural relations. Therefore, this work aims to verify the importance of studying EthnoMathematics, the socio-cultural reality, highlighting the contextualization with an emphasis on culture to rescue popular knowledge. In this sense:

> What I call the EthnoMathematics Program is a research program in the Lakatosian sense that has been growing in repercussion and has shown itself to be a valid alternative for a pedagogical action program. EtnoMatematica proposes an alternative epistemological approach associated with a broader historiography. It starts from reality and arrives, in a natural way, through a cognitive approach with a strong cultural foundation, at pedagogical action (D'AMBROSIO, 1993, p. 6).

D'Ambrosio (1993) used an etymological resource composed of three Greek radicals *ethno, mathema* and *TICs* to explain what he means by EthnoMathematics.

For D'Ambrósio (1985), EthnoMathematics is the mathematics practiced by distinct cultural groups that are identified as indigenous societies, groups of workers, professional classes, groups of children of a certain age, etc.

Thus, D'Ambrosio (1990) defines EthnoMathematics as the study of mathematical ideas and practices that have been developed by specific cultures (ethno vs. ethnic) throughout history, using techniques and ideas (ICTs = technique) appropriate to each cultural context, with the aim of learning to deal with the environment by working with measurements, calculations, inferences, comparisons and classifications. In this way, these cultures have developed the ability to model the natural and social environments, according to their own needs, in order to explain and understand the phenomena (*mathema*) that occur in them.

> Ethno: It is now something very broad, referring to the cultural context and therefore includes considerations such as language, jargon, codes of behavior, myths and symbols; Matema: It is a difficult root, which goes in the direction of explaining, knowing, understanding; Tica: It undoubtedly comes from Tchne, which is the same root as art or technique of explaining, knowing, understanding the various cultural contexts (D'AMBRÓSIO, 1990, p. 5).

EthnoMathematics in Borba's concept (1993, p. 56) was the expression of traits in a culture that seeks to identify its own cultural problems and solutions, thus identifying its generating activities, in other words, "a Mathematical form that expresses traits of a given culture, in an attempt to solve problems that are an expression of this culture".

According to Knijnik, EthnoMathematics investigates practices that develop skills such as decoding knowledge.

> The investigation of the traditions, practices and mathematical conceptions of a subordinate social group (in terms of volume and social, cultural and economic composition) and the pedagogical work to ensure that the group: interprets and decodes its knowledge; acquires the knowledge produced by academic mathematics; establishes comparisons between its academic knowledge, analyzing the power relations involved in the use of these two types of knowledge (KNIJNIK, 2001, p. 88).

Quoted by Ferreira (1997, p. 22), Ascher defines EthnoMathematics as "the mathematics of non-literate peoples, recognizing as mathematical thought notions that in some way correspond to what we have in our culture". As such, the author argues that each people or group uses their mathematics in a different and useful way, passing on their culture from generation to generation.

Paulus Gerdes (1991) speaks of Ethnomathematics as a program in permanent evolution; "perhaps it is provisionally better to speak of an ethnomathematical accent in mathematics education research or of an ethnomathematical movement" (GERDES, 1991, p.32).

According to Scandiuuzi (2003, p. 5) "[...] the mathematics of different socio-cultural groups and proposes a greater appreciation of the informal mathematical concepts constructed by students through their experiences outside the school context".

EthnoMathematics is totally linked to the biography of the discipline, so the knowledge of all mathematical cultures should be used as a tool to understand its concepts from a memorable perspective. The position in which the trend is inserted is of great value if students are to be able to understand the stimuli and grandeur that govern the formation of Mathematics.

2.2 THE EVOLUTION

Currently, there are many goals for researchers who want to take mathematics education to a new level, differentiated by highlighting the mathematics of everyday life, of our daily lives and of activities that have already been forgotten. These goals are driven by EthnoMathematics. In the past, EthnoMathematics

> [...] it meant non-academic and non-systematized mathematics, that is, oral, informal, "spontaneous" and sometimes hidden or frozen mathematics, produced and applied by specific cultural groups (indigenous people, slum dwellers, illiterate people, farmers...). In other words, it would be a very particular way for specific cultural groups to carry out the tasks of classifying, ordering, inferring and modeling (FIORENTINI, 1994, p. 59).

In 1985, Ubiratan D'Ambrosio published *Ethnomathematics and its Place in the History and Pedagogy of Mathematics, an* article of great historical importance to the trend. "These ideas have stimulated the development of this field of research" (Powell & Frankenstein, 1997, p. 13). In 2003, this article was selected for the NCTM book, *Classics in Mathematics Education Research,* because it had a positive and profound influence on research in Mathematics Education. That same year saw the creation of the *International Study Group on Ethnomathematics* (ISGEm), which launched the

Ethnomathematics program internationally.

More recently, its definition was expanded again by D'AMBROSIO (1998), who made an etymological approximation of the word: ETNO, which refers to something very broad, related to the cultural context, thus including language, jargon, codes of behavior, myths and symbols; -MATEMA- which means to explain, to know, to understand; and -TICA, which comes from *techne and* means art or technique. Therefore, it can be said that EthnoMathematics is the technique or art of knowing, explaining, understanding, dealing and living together in the most varied cultural and social contexts (FIORENTINI, 1994).

EthnoMathematics does not exclude mathematical notions produced and brought by generations of thinkers, but interconnects these values with practical concepts (D'AMBROSIO, 2005). In order to work with EthnoMathematics as a pedagogical action, it is essential to "break free from the Eurocentric pattern and seek to understand, within the individual's own cultural context, their thought processes and their ways of explaining, understanding and performing in their reality" (D'AMBROSIO, 2002, p. 11).

Rosa and Orey (2005, p. 6) explain that Mathematical Knowledge in Everyday Life: An Ethnomathematical Approach "the program emerged to confront the taboos that mathematics is a universal body of study, without traditions or cultural roots".

As a result of this idea, education must adapt and reformulate its methodologies in all areas of human knowledge, but especially in Mathematics. Students must understand the processes of transformation of the knowledge they acquire and produce, seeking to generate a revival of popular culture in the course of lessons, starting from their

own. The consequence of this new proposal would be students building their own

knowledge from everyday experiences.

3. THE HISTORY OF THE GARIMPO AND THE GARIMPEIROS IN THE MUNICIPALITY - HOW THEY AROSE AND THEIR SOCIAL AND ECONOMIC IMPACT ON THE REGION

This chapter will look at the history of the municipality of Matupâ, the first settlements, as well as the first garimpos, one of the main labor activities in the region to this day. We also intend to investigate the environmental, social and economic impacts of mining on the town, which has many natural resources in its surroundings. It will also look at how mining processes are carried out, how experience and technology have helped this practice evolve and how the level of education affects the workers, both in terms of the positions they hold in the garimpo and in terms of the difference in salaries.

3.1 MATUPA

Created from the entrepreneurial vision of the shareholders of Colonizadora Agropecuària do Cachimbo, where they proposed a noble destination for the area surplus to the beef cattle project (Fazenda Sao José), contributing, on the one hand, to the occupation of voids characteristic of the Amazon region, the Urban Project for the city of Matupâ was filed with INCRA in March 1984.

Located 700 km from the state capital, at the junction of BR-163 and MT-322 (formerly BR-080). Matupâ was founded on September 19, 1984. The founding is credited to the Ometto family, through Agropecuària do Cachimbo S/A. The name given by the entrepreneurs Matupâ comes from the Tupi language, a word of Amazonian origin which, in short, has two meanings: One scientific, "Dense bush on the banks of rivers and lakes", and the other humanized, "God-blessed bush", expressed the urbanistic pattern to be adopted: a city that responded to the conditions of ecology that was integrated into the natural environment in which the forest and the river were valued and at the same time responded to our traditions of living in the city.

The city is a very special colonization program that was set up out of the need to create a regional center. It has an urban area with basic infrastructure that is completely ready and a rural network with infrastructure that is sufficiently developed for productive activities to begin. It was created to meet the need for support in a large region undergoing a rapid process of development and was planned to allow the industrialization of products in the region itself.

The urban nucleus of Matupà was elevated to the category of district on December 11, 1985, by Law No. 4,937, when it still belonged to the municipality of Colider, through ADECOM (Matupà Community Development Association), the district won its political and administrative emancipation, with Law No. 5,318 of July 4, 1988. 5.317 of July 4, 1988. With the creation of the municipality and the election held in 1988, the Municipal Organic Law was promulgated in 1990, giving legitimacy and representativeness to the elected legislators.

The municipality has a geographic area of 5,152 km^2 , its seat is 300 meters above sea level, its population, according to IBGE/2007 data, is 14,243 inhabitants, and in the 2008 election the municipality had 8,894 voters. It is bordered to the north by the municipality of Guaranta do Norte and part of southern Parà, to the south by the municipality of Peixoto de Azevedo, to the east by Peixoto de Azevedo and to the west by the municipalities of Novo Mundo and Nova Guarita.

3.2 MINING IN THE REGION

Mining has had the importance of marking various periods in Brazilian history and has established Brazil as one of the largest producers of mineral raw materials, in other words, "the country is certainly among the three richest countries in mineral resources, which is equivalent to saying that the country has one of the largest mineral reserves in the world" (FIGUEIREDO, 1984, p. 13).

Despite being designed to be an agricultural town, since its foundation, the strength of Matupà's economy has always been mining. Located just 8 km from Peixoto de Azevedo, which was simply one of the biggest mining booms at the time, the soil around the town was and still is the source of a lot of gold. Ribeiro (1991) points out that the town continues to live off gold mining and, as it neighbors the municipality of Peixoto de Azevedo, Matupà ends up receiving certain reflexes. Peixoto, in turn, suffered the conflicts of a town that produced more than 1,200 kilos of gold a month, most of it underground.

According to the accounts of the town's old inhabitants, the first settlements in the area were set up to encourage agriculture. However, most of the workers who arrived here were seduced by a greater desire, the possibility of striking it rich in mining. The large number of rivers in the region greatly benefited exploration and, as the gold came to light, more and more people began to move to the area and soon the borders of the new town began to emerge, as it became clear that so many families needed to be served commercially, and so the first hospitals, schools, churches, markets, etc. were born.

The prospectors then began their excavations, without any management, experience or care for the dangers to health and the environment, driven only by the search for riches. The mountains were converted into ravines by digging them up to remove the nuggets, which were very rare to find.

The profits and expenses of the gold found remained with the partners, the workers realized that they would earn very little, even though they worked day and night, but this didn't discourage them. Fights over territory occurred frequently, theft of mined gold was also common and, consequently, many people lost their lives in the process.

Another difficulty constantly faced by the workers is illegality. Various actions by the environmental police have taken place over the years, but even so this type of repressive action has not solved the problems at all, as the composition of clandestine mining has always proved to be very dynamic and versatile. Therefore, some authors point to ways forward, such as the formation of a policy to regularize and recognize the rights of mining workers, as a proposal to be considered in order to minimize the conflicts experienced by these men.

Therefore, the history of gold mining in the municipality of Matupâ is made up of hard-working, impetuous and ambitious men, pioneers in opening up forests and exploring for gold, in other words, individuals driven by the dream of one day achieving wealth, in short, the good fortune of prospering, the desire of every gold miner (GUERREIRO, 1984).

Taking into account the importance that gold mining has had and still has for the economic activities of the municipality and the region, we understand that this project is relevant, as it comes to bring to society a little of the mathematical culture of the miners and to bring to the classroom this context of the activities carried out in gold mining. To understand the process from the extraction to the sale of gold, to investigate why this practice, which arises from necessity even in times of scarcity, has become the only source of income for countless families in the north of MT.

3.3 THE PROFESSION

One of the most rustic means of mining is the process of garimpagem, as it is usually identified in remote regions without access to basic resources, and is often not legal. In Brazil, this practice has already had significant economic value and began in colonial times, before and after independence.

The country's great current challenge is to exploit the immense mineral wealth already mapped out in its subsoil in a socially and environmentally sustainable way, without repeating the damage of the past and eliminating clandestine mining. It's not easy, but Parliament and the Federal Government already have at least some mature proposals on the subject. The central idea is to take advantage of the country's mineral potential in a legal way to generate jobs and income for miners and other workers in the area. This strategy is summed up by geologist Clàudio Scliar, former secretary of the Ministry of Mines and Energy, who played an important role in drawing up the National Mining Plan that will be in force until 2030.

"Our concern with the new regulatory framework is to have legal procedures that make it easier to resolve problems in the appropriate instances, so that anyone who wants to invest in research and extraction of mineral goods is supported and has the conditions to have access to these areas in order to better generate income and employment, which is the main objective of using mineral goods."

The name - - came from a pejorative word - Grimpeiro. In the past, the grimpeiros climbed the crags to evade taxes. They were grimpeiros, later garimpeiros. Today the name no longer has a pejorative meaning. It's the name of the men who struggle to extract precious stones or gold from alluvial land or break up gravel in the search for precious metals.

The Mining Code defines a prospector as follows: *Decree-Law No. 227/67, Article 70:*

> The individual work of those who use rudimentary instruments, manual devices or simple, portable machines to extract precious and semi-precious stones and valuable metallic or non-metallic minerals in alluvial deposits, in the hollows of watercourses or on reserved banks, as well as in secondary deposits or on plateaus, slopes and hilltops, deposits which are generically referred to as garimpos.

However, even today, this practice is extremely exploitative. Many garimpeiros work hard and earn little, paying high fees. When they don't have the tools they need to mine or the money to stay in their place of work, they look for all kinds of deals, for example, the **Meia-praça**: when the worker finances and keeps 50% of what is mined. Another form of agreement that is quite common in this area is called the **Subjection System**: the worker takes out the loan and has the right to choose the tools and the place where he will mine, but he will receive and pay the price that the owner of the garimpo imposes.

Mining has always been an exercise full of contradictions and disagreements, because on the one hand it contributes to the country's economy and helps to spread its territory, and on the other it damages the environment and poses various health risks.

The use of mercury, for example, used in some mining operations, has proved to be one of the biggest problems for nature, as it is toxic and can contaminate workers, rivers, fish, wild animals and people who use the region's waters. In the event of possible poisoning, when mercury is ingested or inhaled, it is immediately absorbed into the bloodstream via the lungs and can cause, among many other problems, nervous system deficiencies and loss of motor coordination.

Despite the difficulties and risks faced, it is in garimpo that many men make a living for their families. Although difficult and risky, it is one of the only professions available to those with little schooling and who live in remote areas.

3.4 PROGRESS

Although garimpeiros suffer from the working conditions nowadays, in the past, without any resources or technology, it was much worse. Everything was extremely precarious; the heat was harsh; the tools they built were damaged and unsafe, and the ravines were high. However, despite the primitive conditions, the miners risked day and night digging in the hope of getting rich.

Alluvial gold is still mined today using bateias, a kind of basin made of aluminum or wood, similar in shape to a deep dish. Water is added and, with a circular motion, the water is removed and the gold settles at the bottom of the container.

The most common tools used in small garimpos today are: shovels, sieves, small canoes, agitators, bicams (an artificial wooden bed that diverts the flow of a river or stream to allow mining), etc. In larger areas, where the owner has more financial capital, they also have modern and efficient machinery such as metal detectors, loaders, conveyors, tractors, etc.; in this case, the miners must have completed their schooling, as it is necessary to master the handling of such machinery.

Although the purpose of gold mining remains the same, new techniques have offered workers more efficient technologies for the practice:

Sluice gates: The use of "sluices" takes advantage of the principles of filtration, but uses mechanization and a much larger capacity in order to make it economically viable on a large scale. Earth and sand are loaded into a sluice box, a long metal blade with grooves at the bottom. Water is then added to the mix. While the heavier gold settles on the ridges, the lighter material is carried out of the sluice box. The gold can be recovered.

S **Underground mining:** Although more laborious and dangerous than other modern gold mining techniques, underground mining is a popular and lucrative technique. Although the mining of precious minerals has taken place since prehistoric times, it is still a popular form of gold mining. Mechanized mining equipment and safety regulations make modern underground mining safer and more efficient than in previous years.

S **Gold** cyanidation: This process involves adding the toxic chemical cyanide to rocks that are likely to contain small amounts of gold. The gold nuggets with cyanide can then be separated from the rock in which they were found. This combination is called cyanide gold. Zinc is added to the cyanide gold in order to remove the cyanide itself, then sulphuric acid is added to the zinc/cyanide mixture in order to remove the zinc. What remains is a paste of pure gold ore.

S **Metal detection:** Modern metal detectors can be used to find gold, but they don't distinguish between it and other metals. And the fact that a tin can is much larger than the average-sized gold nugget you're likely to find means that other, less valuable ores will show up on a metal detector much more often than gold.

Researchers around the world are frequently studying and perfecting new low-cost, non-toxic techniques that can be used on a large scale to minimize the damage that gold mining causes to the environment. In the meantime, however, nature is susceptible to all kinds of devastation in this regard.

3.5 THE STRATEGIES

There is a need for financial and economic support for garimpo activity, as it is a major source of income that balances the country's growth rates to a very significant degree. This is especially the case when you consider the potential of Brazilian soil, which has very atypical and rich configurations, capable of placing garimpo at very balanced levels with regard to the development of stratification.

Until 1998, the miner retired as a special insured person, based on a contribution of 2.1% of the result of the commercialization of production. In this case, the worker doesn't need to prove that they have paid contributions, just that they have worked for years. Since Constitutional Amendment 20/98, miners have been classified as individual taxpayers and must pay 20% of their income. Most garimpeiros, however, do not pay this tax.

The fact is that the income from such hard work ends up in the hands of a few. The garimpeiros who work day and night in a dangerous and exhausting activity make up a rather helpless group, most of whom have no formal education. The profits remain in the garimpo itself, as everything is charged for: food, clothes, drinks. The currency that circulates is gold.

In this study, when interviewing a small group of garimpeiros, we were able to highlight the huge difference in salaries due to their level of education. Those who have no formal knowledge are offered the rudest jobs, while those who have completed high school as well as a qualification can benefit from a better position with an equivalent monthly income.

As a result, some workers reported that they were now attending school. In the search for a better job and salary, they are looking for the municipality's Youth and Adult Education Center (CEJA), where they can complete their studies at times that suit their work routines.

What was fascinating, however, was to detect the mathematical knowledge acquired by these workers as a result of their profession. Some of them, even with a low level of education, are able to perform percentage calculations perfectly, and report that it was necessary to learn in order to avoid being cheated out of their wages.

According to the reports, it can be concluded that the trades generated by gold in the municipality, directly or indirectly, provide a livelihood and allow men with few qualifications and no professional training to work informally. In addition, the need has encouraged gold diggers to return to school, as they can improve their knowledge and have access to their rights as citizens.

4. THE SCHOOL - THE ETHNOMATHEMATICAL KNOWLEDGE OF THE MINERS OF MATUPA

This topic tells how the research was introduced into the school; how the students received the proposal and the results of this experience, which took them from theory to practice. It highlights the link between the cultural world of opinions, concepts and knowledge of the group studied and the world of regular learning developed in the school environment. We adopted this methodology because we believe that it is possible to add popular knowledge to systematized knowledge in order to consolidate more meaningful knowledge from the perspective of EthnoMathematics.

4.1 THE RESEARCH

The research is qualitative, with a bibliographic approach and field research. It began with searches in books and websites to learn more about the concepts of EthnoMathematics, and to find out more about the subject in relation to the specific group of miners.

With this material in hand, a pedagogical proposal was drawn up with the aim of bringing to the classroom the mathematical content used by the gold diggers in carrying out the activities of extracting and selling gold, such as percentages, for example, which made it possible to arouse interest in an activity that was in line with the school syllabus and at the same time was different because it dealt with Cultural Mathematics.

After applying some mathematical concepts in the classroom, a small survey was carried out in order to identify how many students in the class had someone in their family who worked in the mining industry so that they could carry out the interviews, given that it would be risky to take them to a mining site due to its hostile environment. Nine students said they had someone at home who made a living from the mining process, and after being instructed they were given the task of interviewing this family member.

The purpose of the interviews was to gather data on the mathematical knowledge involved in the daily lives of the miners, to understand and interpret certain behaviors, opinions and expectations of this group. They were exploratory in nature, so the aim was not to obtain numbers as results.

Although they were about to conduct the interviews with their own relatives, some of the students were not fully prepared for this task, so it was essential to create a communication script, with a few rehearsals among classmates, to relax and stimulate the students. Since there were few activities of this kind, they were also given guidance on how to talk to the interviewees, how to present the proposal and how to thank them for their attention.

In order to improve the research work, the investigative look of the interviews was able to add details that enriched the students' teaching and learning. The use of tape recorders and pre-structured questions proved to be effective in polishing the large volume of raw information received and interpreting it in the best possible way, but it wasn't possible to use them in every conversation because some miners were embarrassed to answer using them. They then answered the questions by hand.

4.2 THE RESULTS

Lessons that make students understand that everything in mathematics will one day be applied to the real world will bring back the most dedicated and interested students. We need to prepare them for life and help them see that mathematics will lead them to a better understanding of reality, to organize their reasoning and to make countless discoveries.

With these points of view in mind, the classroom project began with the presentation of the central idea along with some brief concepts of the contents of Percentage, Area and Volume of flat figures. These concepts were expected to be present in the professional lives of the miners. Figure 1 shows the students in their first contact with the study, where they were separated into groups to carry out some exercises.

Photograph 1 - Students carrying out the first activities Source: Fabiana A. Leite, 2018.

As the picture shows, the students were very excited about the project. They took several *selfies,* all at the request of themselves and the principal, which were posted on the school's *Facebook page.* The activities were carried out successfully, with only a few doubts arising during the lesson. All these were solved with the help of the class teacher.

Some of the activities carried out by the students were as follows:

1) Calculate and record the results:

a)   10% of 200

b)   10% of 500

c)   10% of 80

d)   5% of 200

e)   50% of 500

f)   20% of 80

g)   15% of 200

h)   25% of 500

i)   5% of 80

j)   30% of 200

k)   75% of 500

l)   35% of 80

Photograph 2 - **Student notebook** Source: Fabiana A. Leite, 2018.

2) A pizza has been divided into 8 equal pieces.

a)   If you ate 4 pieces of this pizza, what percentage would you have eaten?

b)   What if you ate 2 pieces?

c)   If you had eaten 6 pieces, what percentage of the pizza would that be?

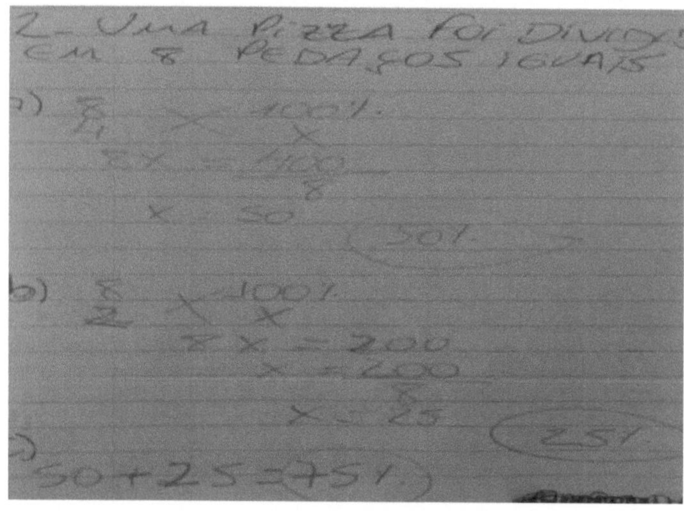

Photograph 3 - **student notebook** Source: Fabiana A. Leite, 2018.

3) Outline the amount corresponding to 35% of the following colored pencils:

Photograph 4 - student notebook Source: Fabiana A. Leite, 2018.

4) Calculate the area of a rectangle whose perimeter is 72cm and height is triple the base.

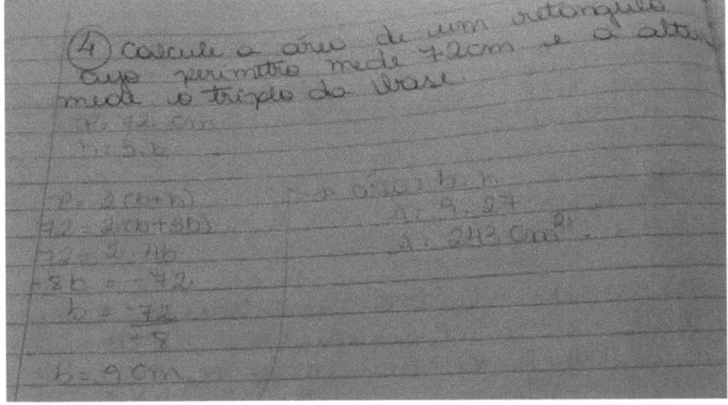

Photograph 5 - student notebook Source: Fabiana A. Leite, 2018.

5) Calculate the volume of the solids below:

a)   Cube with an area of 3cm.

b)   A parallelepiped with a base of 5m, a length of 3m and a height of 2m.

c)   A cylinder with a base of 3m and a height of 10m.

d)   A pyramid with a base of 3m and a height of 10m.

e)   Sphere of radius 2m.

f)   Cone with base 1m and height 5m.

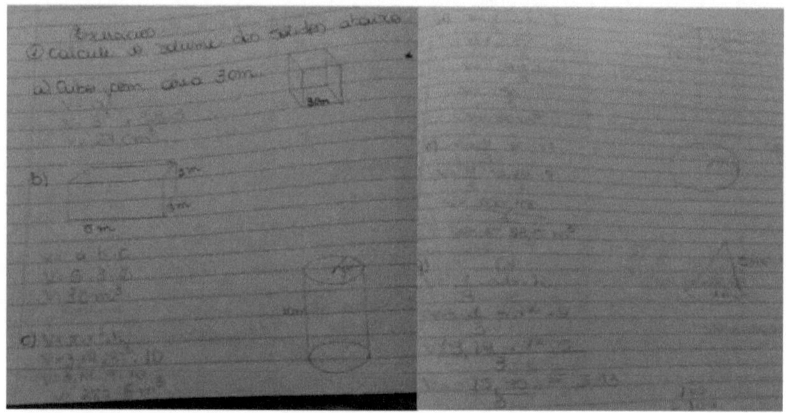

Photograph 6 - Student notebook Source: Fabiana A. Leite, 2018.

After seeing and reviewing these concepts, nine of the students in the class set off for the interviews. Among the many questions asked in order to have a meaningful informal conversation, we would like to highlight those that relate to the conception and development of this research, as well as the interviewees' consecutive responses: The first question was to identify the age group of the interviewees.

Graph 1 - Age group of garimpeiros.

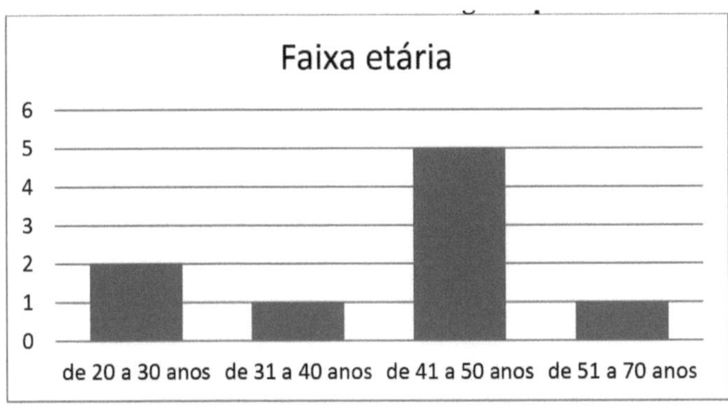

Source: Fabiana A. Leite, 2018.

Graph 1 shows that the profession is not just the choice of older men. The age range of those interviewed varies from 19 to 69, and the fact that young people are also engaged in this exercise is reflected in graph 2, where it is possible to see a high rate of workers attending school.

Graph 2 - Garimpeiros' level of education

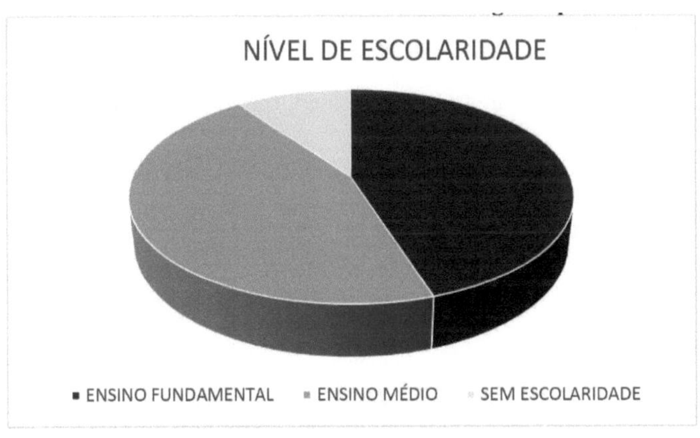

Source: Fabiana A. Leite, 2018.

The second question was designed to find out how long these workers have been in the profession:

Graph 3 - Garimpeiros' years in the profession

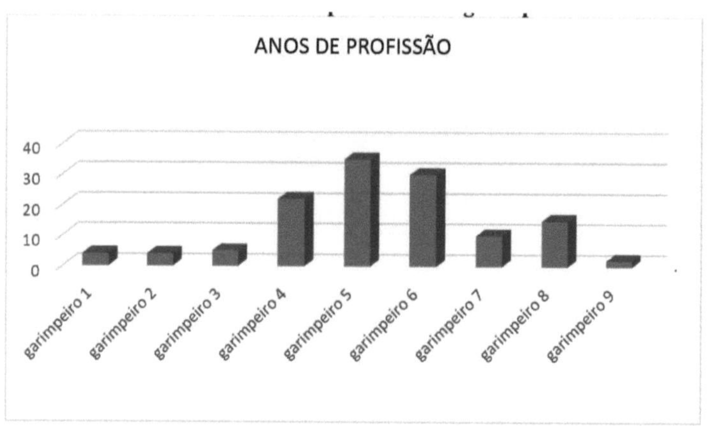

Source: Fabiana A. Leite, 2018.

Graph 3 shows a greater variation in the length of time the workers have been working in the profession. Of particular note is garimpeiro 5, who reported that he had been working for more than 30 years.

When asked how the excavation measures are determined, the majority were unable to answer, as they do not work in this sector. The others said that the wells are measured and dug using machines. Using them, it is possible to obtain tracks with precise measurements for mining. Only one of the interviewees reported working with these machines. Garimpeiro **i**:

> [...] We do a soil survey to see if there's gold there, then we send it to the owner of the garimpo to see if it's worth digging there. Then, when he authorizes it, I, the tractor driver, go with the excavator to dig the hole, which has to be 8m deep, it can't be just any size.

They were also asked whether their salary was fixed or a percentage of the gold mined. Unanimously, they all said they received a certain percentage of the gold mined. These are the amounts they receive:

Graph 5 - Percentage of garimpeiros' salaries

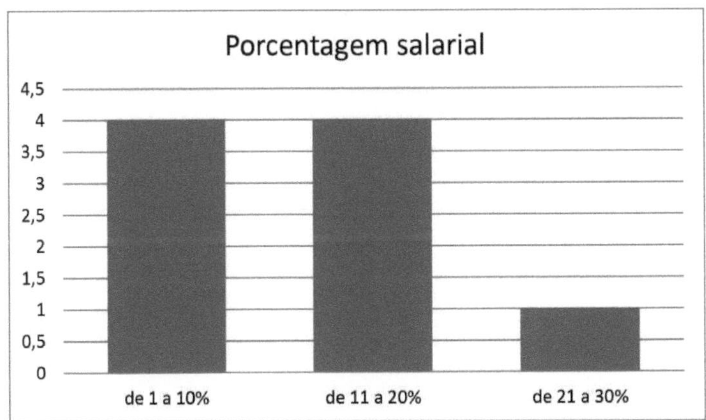

Source: Fabiana A. Leite, 2018.

Graph 5 shows how big the wage gap is between the workers. By analyzing the interviews, it was possible to see that the garimpeiros who have completed high school are the ones who earn a higher percentage of their salaries, and those who earn less are already looking for school in order to finish their studies and gain a better position and salary. Garimpeiro **h**:

> [...] today I receive 15% but before I received less because I didn't know how to count. I joined CEJA two years ago to finish my studies and now I know how to count percentages.

One observation to be made on this question is that everyone reported knowing how to calculate percentages, even those who didn't have a complete education, who said they had learned so as not to feel cheated when they were paid. Garimpeiro **f**:

> [...] I've always worked in the mines since I was a kid and I never had the chance to finish my studies, but I know exactly how much I have to earn (...) working in the mines I had to learn the accounts.

The interviewees were also asked if they knew how many grams of gold are currently removed in the garimpo where each one works, all of them said yes and gave the following figures:

Graph 6 - Grams of gold found daily

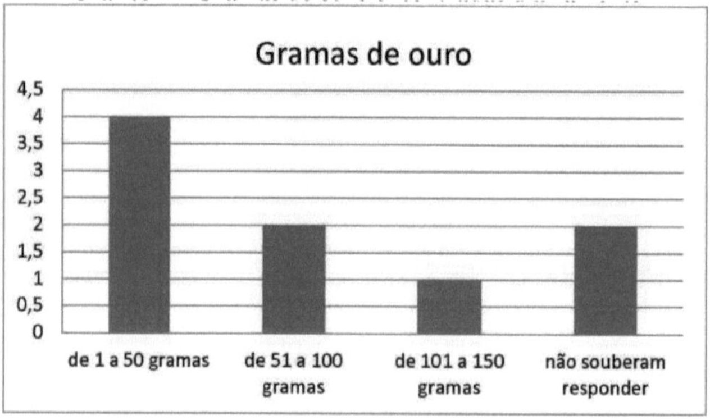

Source: Fabiana A. Leite, 2018.

Despite the fact that the municipality has had garimpos since it was founded, there is still a lot of gold to be found in the region, and graph 6 shows this perfectly. Taking into account that the price per gram is currently coming out at R$138.26, in a single day a garimpo makes a profit of more than R$4,100.00.

Question number six of the questionnaire was intended to find out which techniques are used today in the respective garimpos. Only garimpeiro **e** said that he carries out the process manually, all the others reported that the mining method is done using machines. Garimpeiro **g**:

> [...] the boss has four backhoes that are used to dig the first holes. Then we
> use a water pump to break up the ravines that have gold dust or nuggets.
> There is almost nothing manual there, all the "pioes" work with some kind
> of machine.

The student who interviewed garimpeiro **g** provided some photos of the aforementioned garimpo showing garimpeiros dismantling ravines with water jets, as shown in figure 7.

Photo 7 - Garimpeiros with water jets

Source: Fabiana A. Leite, 2018.

The next question asked if they could tell the length, width and depth of the excavation tracks. Of all the interviewees, 4 were unable to answer the above question. The others said that on average the excavation tracks are 20m long, 60m wide and 8m deep. Garimpeiro **i**:

> [...] as I said, the depth is 8 meters, then the width is about 60 meters and
> the length is 20 meters. On average, it's that size. Sometimes it can vary
> depending on the location, but it's almost always that size.

Figure 8 shows an example of the magnitude of a track being excavated, in the background you can see a backhoe, a machine that was mentioned a lot in all the conversations we had.

Photo 8 - Track being built

Source: Fabiana A. Leite, 2018.

The last two questions of the interview asked about the costs of maintaining the garimpo, such as the kitchen service and broken equipment. All the interviewees reported that the cook's salary or any defects in the machines or objects used for mining are borne exclusively by the owner of the garimpo. Garimpeiro **d**:

> [...] it's the owner who buys the food and pays for the garimpo cook [...] and if the tractor or one of the pumps breaks down or is faulty, it's the garimpo owner who pays for it too, our job is to mine, if it breaks down it's up to him.

All the dialogues were successfully concluded, and the answers obtained showed that the workers felt comfortable answering every question put to them. We also realized how rich their knowledge is, a class that works in difficult conditions but has never given up on the profession.

After collecting all the interviews in class, each one was analyzed together with the class. When asked how well the research was going, one of the students in the role of interviewer said that his father felt important to know that someone was researching his profession.

The main question posed to the students in the analysis of the dialogues was the guiding question of this project: what mathematical knowledge do the prospectors have and use in their gold mining activities in the Matupâ region? By listening to each audio and reading each handwritten answer, they concluded that the prospectors have notions of:

Addition;

S subtraction;

Multiplication;

Percentage;

S length measurements;

S mass measurements;

Area measurements;

Volume measurements;

Time measurements;

S angle measures;

S geometries.

With this analysis, in addition to the results relating to the calculations, the students understood mathematics in its practical form, as an answer to that age-old question they always ask themselves: "Professor, where are we going to use this calculation in our lives?". After seeing that their parents and relatives, even those with less education, were able to carry out these mathematical operations, they felt motivated and much more involved with the subject.

We believe that, through this pedagogical proposal, extraordinary transformations were achieved in relation to teaching and learning in the subject of Mathematics, because this activity created an extremely important connection between school Mathematics and the Mathematics contained in their realities.

5. FINAL CONSIDERATIONS

The aim of this work was to investigate what mathematical knowledge the workers who survive by mining in the municipality of Matupà have, and what they use in their profession. In this approach, this data was presented in order to provide an analysis of this little-explored universe of mathematics teaching, with the aim of identifying and applying this trend and paving the way for the subsequent enhancement of cultural collections or, at the very least, laying the foundations for further research.

Through this research project we have seen that, in a specific group of workers, more than half do not have completed their schooling and, even so, they all show certain mastery and notions of Mathematics and, when we talk about this knowledge, it does not mean that they can perform all the calculations with formulas and theoretical concepts. Through an EthnoMathematical survey, it can be seen that they have notions of organization, comparison, quantification, and this is the most relevant information in the survey results.

One of the observations gained from analyzing the interviews is that, in general, the miners are unaware of a large part of formal mathematics, the so-called conventional mathematics learned at school. However, through their profession they have acquired incredible knowledge and know how to use it accurately in their daily lives.

By bringing all this information into the classroom, the pedagogical proposal was to make mathematics live, to encourage students to deal with real situations that are part of their lives outside the school environment, i.e. family life. By doing this, it was possible to delve into cultural roots and, as well as contributing to the school, to value this transcultural or transdisciplinary tradition.

From this point of view, the circumstances experienced by the students have led to the formation of mathematical knowledge, and EthnoMathematics comes to provide progress in the ability to learn this field of human knowledge accurately. Remembering Piaget, we must bow to this need if we want to train individuals who are capable of creating, producing, discovering and not just repeating.

6. REFERENCES

BORBA, M.C. EtnoMatemàtica e a cultura da sala de aula. **Revista da sociedade brasileira de educaçâo Matemàtica**. N.1 (2nd semester), 1993.

BRASIL, **Parâmetros curriculares nacionais**: Matemàtica / Ministry of Education. Secretariat for Primary Education. 3.ed. Brasilia: The Secretariat, 2001.

D'AMBROSIO, U. (1985). Ethnomathematics and its place in the history and pedagogy of mathematics. **For the Learning of Mathematics, 5(1)**, Bristol, UK, Laurinda Brown (Ed.), p. 44 - 48.

D'AMBRÓSIO, U. **EthnoMathematics. The** link between tradition and modernity. 2^a Edition. Belo Horizonte: Autêntica, 2002. 110 p. (Coleçâo Tendências em Educaçâo Matemàtica).

. **EthnoMathematics:** a Program. A Educaçâo Matemàtica em Revista.(SBEM), Ano 1, N°1, 2° semestre, 1993, p. 5 - 11.

. (1993). **EthnoMathematics**: a program. A Educaçâo Matemàtica em Revista, 1(1). Blumenau, SC: SBEM, p. 5 - 11.

. (1985). **Ethnomathematics** and its place in the history and pedagogy of mathematics. For the Learning of Mathematics, 5(1), Bristol, UK, Laurinda Brown (Ed.), p. 44 - 48.

. (1990). **EthnoMathematics.** Sâo Paulo, SP: Âtica.

. **EthnoMathematics**. 5th edition, Sâo Paulo: Âtica, 1998.

. **EtnoMatemàtica: link between traditions and modernity**. 2. ed. Belo Horizonte:

Autêntica, 2005a.

FERREIRA, E. S. **EtnoMatemàtica**: Uma proposta metodològica. Rio de Janeiro: Santa Ursula University, 1997.

FIGUEIREDO, Bernardino. **Garimpo and mining in Brazil**. In: ROCHA, Gerôncio (Org.). Em busca do ouro: garimpos e garimpeiros no Brasil. Rio de Janeiro: Marco Zero, 1984.

FIORENTINI, D. **Directions of Brazilian research in Mathematics Education: the case of scientific production in Postgraduate Courses**. Doctoral thesis. Faculty of Education, UNICAMP, Campinas, SP: 1994.

GUERDES, P. **EthnoMathematics**: culture, mathematics, education. Maputo [Mozambique]: Instituto Superior Pedagógico, 1991.

GUERREIRO, Gabriel. **Gold mining in the Amazon**: economic, social and political repercussions. In: ROCHA, Gerôncio (Org.). Em Busca do Ouro: Garimpos e Garimpeiros no Brasil. Rio de Janeiro: Marco Zero, 1984.

KNIJNIK, G. Mathematics education, social exclusion and the politics of knowledge. **Bulletin of Mathematics Education**. Year 14, n.16, 2001.

POWELL, A. B.; FRANKENSTEIN, M. (1997). Ethnomathematical knowledge. In Powell, A. B. & Frankenstein, M (Eds.), **Ethnomathematics: Challenging eurocentrism in mathematics education**. New York, NY: SUNY, p. 5 - 13.

SCANDIUUZI, Pedro Paulo. **EthnoMathematics and the training of mathematics educators.** Available at: <http://www.ethnomath.org/resources/brazil/a-

tnomatematica.pdf>. Accessed on: May 20, 2018.

45

I want morebooks!

Buy your books fast and straightforward online - at one of world's fastest growing online book stores! Environmentally sound due to Print-on-Demand technologies.

Buy your books online at
www.morebooks.shop

Kaufen Sie Ihre Bücher schnell und unkompliziert online – auf einer der am schnellsten wachsenden Buchhandelsplattformen weltweit! Dank Print-On-Demand umwelt- und ressourcenschonend produziert.

Bücher schneller online kaufen
www.morebooks.shop

Printed by Books on Demand GmbH, Norderstedt / Germany